I0817433

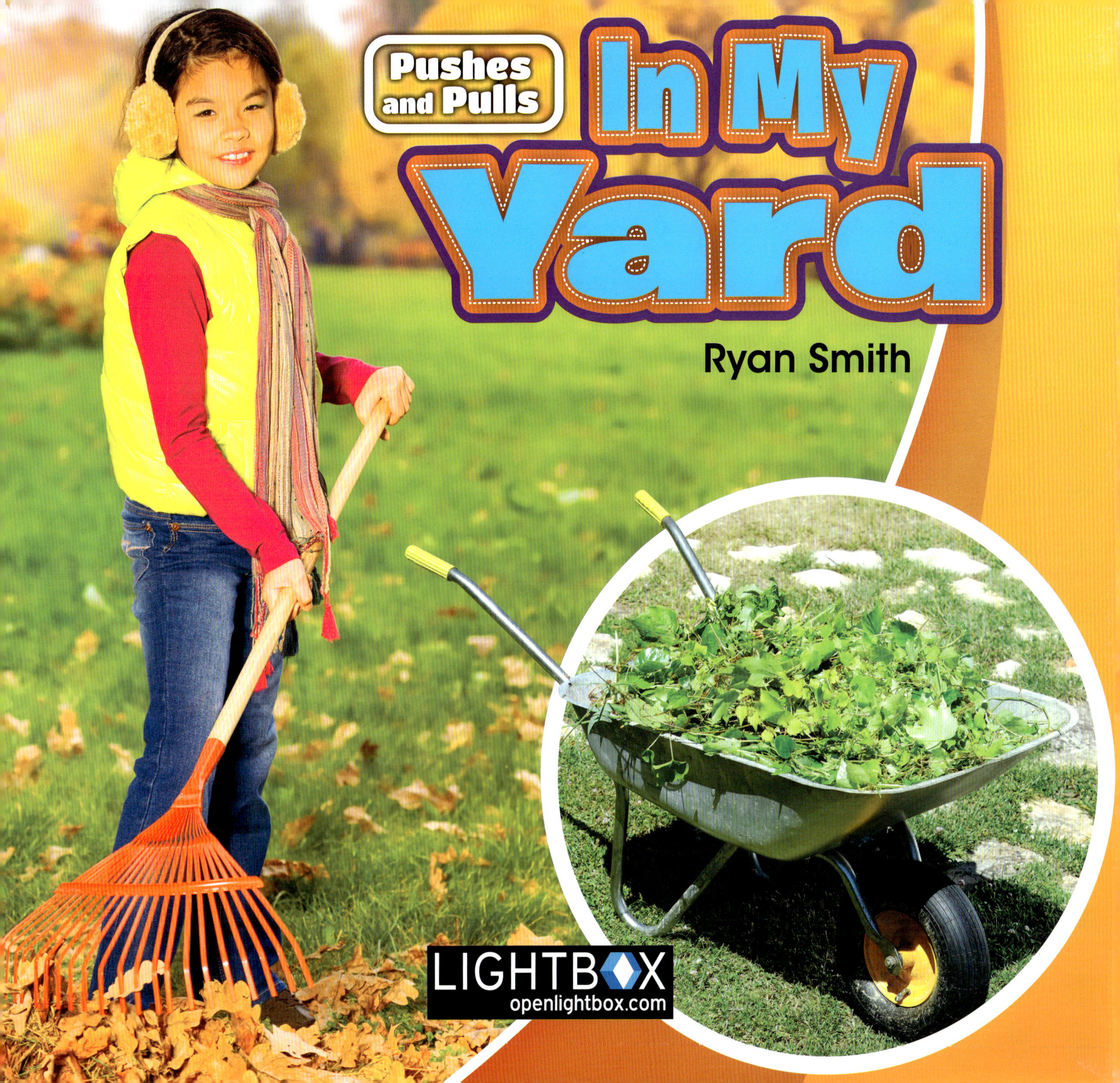
Pushes and Pulls
In My Yard
Ryan Smith
LIGHTBOX
openlightbox.com

LIGHTBOX

Go to
www.openlightbox.com
and enter this book's
unique code.

ACCESS CODE

LBXC7552

Lightbox is an all-inclusive digital solution for the teaching and learning of curriculum topics in an original, groundbreaking way. Lightbox is based on National Curriculum Standards.

OPTIMIZED FOR

- ✓ **TABLETS**
- ✓ **WHITEBOARDS**
- ✓ **COMPUTERS**
- ✓ **AND MUCH MORE!**

STANDARD FEATURES OF LIGHTBOX

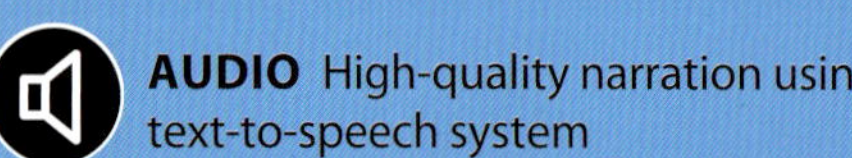

AUDIO High-quality narration using text-to-speech system

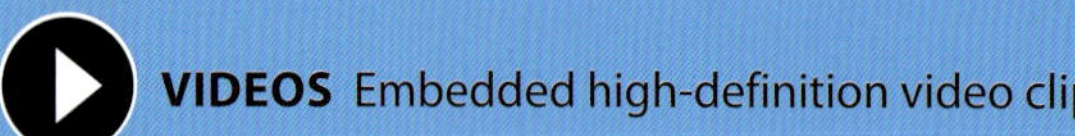

VIDEOS Embedded high-definition video clips

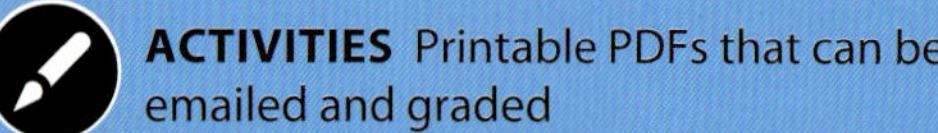

ACTIVITIES Printable PDFs that can be emailed and graded

WEBLINKS Curated links to external, child-safe resources

SLIDESHOWS Pictorial overviews of key concepts

INTERACTIVE MAPS Interactive maps and aerial satellite imagery

QUIZZES Ten multiple choice questions that are automatically graded and emailed for teacher assessment

KEY WORDS Matching key concepts to their definitions

VIDEOS

WEBLINKS

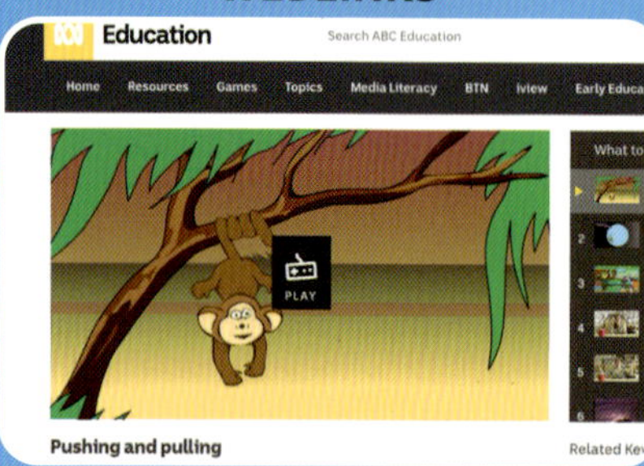

SLIDESHOWS

QUIZZES

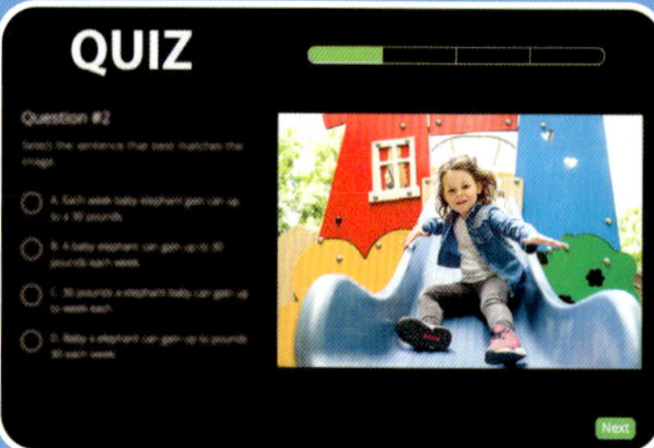

Contents

We push and pull things every day. We push things to move them away from us. We pull things to move them closer.

I am helping my family in the yard today.

We can learn about the science of pushes and pulls while we work.

First, we clean up the leaves.

I use a rake to pull the leaves into a pile.

My brother wants to make the pile taller.

How can I make the pile taller using my rake?
People have been using rakes for 10,000 years.

We have to pull weeds next. I grab the top of a weed and pull.

The weed is hard to pull out.

My mom shows me how to grab the bottom of the next weed and pull. It is easier to pull this one.

How can I make pulling the weeds even easier?

My sister has filled the wheelbarrow with things for the compost bin.

I push the wheelbarrow. It rolls forward.

Most of the **wheelbarrows** in the **United States** are made in **Pennsylvania**.

I push the wheelbarrow to the compost bin.

What will happen if the wheelbarrow runs into the compost bin?

Our last job is to water the garden.

I start to pull the hose.
It is heavy.

My brother says I should pull the end of the hose.

Why is it easier to pull the end of the hose?

A **garden hose** can weigh as much as **a bowling ball.**

See what you have learned about pushes and pulls in the yard.

What do you push in the yard?
What do you pull in the yard?

KEY WORDS

Research has shown that as much as 65 percent of all written material published in English is made up of 300 words. These 300 words cannot be taught using pictures or learned by sounding them out. They must be recognized by sight. This book contains 67 common sight words to help young readers improve their reading fluency and comprehension. This book also teaches young readers several important content words, such as proper nouns. These words are paired with pictures to aid in learning and improve understanding.

Page	Sight Words First Appearance
4	am, and, away, day, every, family, from, I, in, move, my, the, them, things, to, us, we
5	about, can, learn, of, while, work
6	first, leaves, up
7	a, into, use
8	make, wants
9	been, for, have, how, people, years
10	next
11	hard, is, out
12	it, me, one, shows, this
13	even
14	has, with
15	are, made, most, states
17	if, runs, what, will
18	last, our, water
19	start
20	end, says, should, why
21	as, much

Page	Content Words First Appearance
4	yard
5	pulls, pushes, science
7	pile, rake
8	brother
10	top, weeds
12	bottom, mom
14	compost bin, sister, wheelbarrow
15	Pennsylvania, United States
18	garden, job
19	hose
21	bowling ball

Published by Smartbook Media Inc.
14 Penn Plaza, 9th Floor New York, NY 10122
Website: www.openlightbox.com

Library of Congress Control Number: 2020014796

ISBN 978-1-5105-5460-3 (hardcover)
ISBN 978-1-5105-5461-0 (multi-user eBook)

Printed in Guangzhou, China
1 2 3 4 5 6 7 8 9 0 24 23 22 21 20

052020
110819

Designer: Ana María Vidal Project Coordinator: Ryan Smith

The publisher acknowledges Getty Images, iStock, and Shutterstock as the primary image suppliers for this title.